Breakfast at Rudy's

Moving from Grief to Grace

REBECCA COLLISON

ISBN: 978-1-967375-30-1 (Paperback)
ISBN: 978-1-967375-31-8 (E-book)

Library of Congress Control Number: 2025913581

Printed in the United States of America

Published by:

info@thequippyquill.com
(302) 295-2278

To those who saw beyond the ordinary and
challenged me to do the same.

Table of Contents

Introduction

This book is my third view from within the fields of my life.

My first book, Preparing Fields in Seasons of Change, is written from my academic self, a book created from my doctoral research at Wesley Theological Seminary, Washington, D.C. At the encouragement of a professor at the school, I wrote the book and Trilogy Publishing printed it. However, by the time the printed book came into my hands, we were in the 2020 COVID pandemic lockdown.

My second book, Groundwork: Farm Parables and the Cultivation of Faith, is written from my pastoral self, a book to extend the learning from my research into practical application within the local church. To connect Jesus' parables of the land to one's own life and the world around them, I wrote this book in 2021 in a format that could be used as a sermon series or small group study. With life changing so fast, I re-published the book in 2025.

"Breakfast at Rudy's: Letters to God in search of life in unexpected places after loss is written from me while on a spiritual renewal leave. As I began the renewal leave, all I could see were the problems in front of me, knowing God had a plan, but failing to see it. Loss and trials from the previous four years prompted the book, but the Holy Spirit unfolded exactly how the book would read: as letters to God recounting a difficult season and the learning to love and live fully once more. By the end of the leave, the promise was before me, and hope was clearer than ever before. Borne of conversations at the table.

The image of the table is interwoven in the book–breaking bread with others, reclaiming an appetite for life, finding hope in the experience. Throughout Scripture, the table represents an intersection of the daily and the divine, the scars and the sacred, the hurts and the hope. While Rudy's is an actual restaurant, it stands as a symbol of the ordinary places where the Living Christ meets us and teaches us about the abundant life, we can all have today. In every season, the ordinary places meet unwanted interruptions to show us the Jesus Christ we often forget to see.

This book is a compilation of reflections, prayers, and journal entries during my time of reclaiming my appetite for life, to once again "taste and see that the Lord is good." The "love letter from God" is based on what I heard during my prayer time in those times of sweet, silent solitude. And like a diner leaves a tip, I leave a blessing for all who journey with me in these pages.

With the encouragement of friends and family, I compiled my spiritual and emotional journey on these pages. I hope and pray that I did what was honorable to God.

A WORD IN SEASON I

Now that same day two of them were going to a village called Emmaus, about seven miles from Jerusalem They were talking with each other about everything that had happened As they talked and discussed these things with each other, Jesus himself came up and walked along with them; but they were kept from recognizing him.

He asked them, "What are you discussing together as you walk along?"

They stood still, their faces downcast One of them, named Cleopas, asked him,

"Are you the only one visiting Jerusalem who does not know the things that have happened there in these days?"

"What things?" he asked.

"About Jesus of Nazareth," they replied. "He was a prophet, powerful in word and deed before God and all the people. The chief priests and our rulers handed him over to be sentenced to death, and they crucified him.

***But we had hoped** that he was the one who was going to redeem Israel. And what is more, it is the third day since all this took place. In addition, some of our women amazed us. They went to the tomb early this morning but didn't find his body. They came and told us that they had seen a vision of angels, who said he was alive. Then some of our companions went to the tomb and found it just as the women had said, but they did not see Jesus.*

Luke 24:13-25

3

But we had hoped, but I had hoped... that is where many stories of loss and brokenness begin.

I had hoped to have a little more time with my loved one.

I had hoped the situation would turn out differently.

I had hoped I would be accepted for who I am, not rejected.

I had hoped I could trust that person, but instead was betrayed and abused.

Walking on the road of life means, at times, the solid paving stones will feel like shifting sand beneath our feet. Through the landscape of ups and downs, travelers carry heartbreak and hurt of lost hope like the two on the Road to Emmaus.

And as Jesus joins us, we are oblivious to his presence. Even in the presence of the Scriptures that affirm the persistence and presence borne out of the Love of an Almighty God, we can keep looking down. Losing sight of the way. Losing sight of the truth. Losing sight of the light of hope.

In the gospel of Luke, two travelers – one named and the other anonymous – are sharing their thoughts on recent accounts in Jerusalem. Cleopas' name is given because we all know a Cleopas, a person in the community who is known to all. Cleopas, who "stood still, looking sad," speaks to our known self. The outside of our lives that everyone sees. Cleopas is the outwardly known part of each of us who asks the questions, "Don't you know…" about what was done with Jesus.

And then there is the unnamed traveler. Many of the Gospel accounts point to a person who is identified, not by name, but by circumstance or societal label. No name is given because we all know that person who is present in the crowd but is silent. The unnamed traveler speaks to the voice inside our heads and hearts, the voice that we alone hear. A voice that is also asking questions, evaluating situations, and forming answers, all inside our minds. Answers that may or may not be correct, but remain hidden in our hearts.

And along comes Jesus, who does not disclose his identity as the Risen Lord. At least not yet. He begins by asking what they are discussing so intently, and upon receiving their answer, answers by opening the Scriptures. The Word of God, from the words of Moses and the prophets, foretold that the Son of God, full of truth and grace, would come to set them free. Free from their sin. Free from feeling lost. Free to embrace a new reality. But first, reality has to be acknowledged.

For it was only when a stranger joined them on their seven-mile journey from Jerusalem to Emmaus did their grasp of what they perceived as an ending rise to the surface. An ending that was simply a new beginning in disguise.

We, as sojourners in the current world as we know it, know what grief and loss feel like. Emotions that show on the outside, and emotions we feel acutely on the inside. Emotions that, coupled with what we think we know about a situation, can block our view of the hope that is right in front of us.

By the end of the journey, hope arises. Hope is offered to all travelers, including you and me.

Still, the journey is necessary.

Healing is often a process. A process that happens in the ordinary. A process that happens in a community. Healing happens in divine interruptions, sometimes not recognized until after the fact.

And healing happens at the table, where we break bread with another.

Like the two on the Road to Emmaus, the journey to healing starts by being authentic in hearts and minds by putting our questions, our struggles, and our grief before the One who is willing to first listen.

THE LANDSCAPE OF LOSS

Farmers are some of the most courageous people I know.

They put seeds into the ground and then trust powers beyond themselves for life to happen. Those seeds then go on to be life-giving food for others, including us. And yet we don't think about that part. To the extent, I have heard more than one grade school child in the current culture hold the concept that food comes from a grocery store, and when asked where it was before the grocery store, they are clueless

In an age where the proud focus of eating establishments seems to be offering farm-to-table cuisine– that is, sourcing ingredients for a meal locally – we have forgotten we, too, are part of that landscape. Forgotten is the fact that before the food appears on the plate, something had to be broken before having the ability to bear fruit. The same can be said for you and me.

During a difficult season in my own life, I understood what being broken physically, emotionally, and spiritually meant. And not so much like the biblical character Job, who had a really bad day that got worse. The writer of Job's accounts then spent the next 42 pages explaining how this once-happy married man went from having it all to losing all his children, property, livelihood, and wealth. Almost all that is, his faith and his wife remained through the season that seemed to break him. I am thankful my life is not that of Job, but there has been a distinctive breakage in my rose-colored glasses of how I thought life should go.

From December 2020 to December 2024, I faced challenges marked by loss in my life that seemed to pile up upon each other. During this time, my first book, "Preparing Fields in Seasons of Change" (2020), and my second book, "Groundwork: Farm Parables and the Cultivation of Faith" (2021,2024), were published. While my first book has a deeply theological and academic focus, the second book was a pastoral response to the first book, offering Bible Study and reflection that paired with personal and social application. Then the vision of this book, "Breakfast at Rudy's," emerged as a deeply personal journal and revelation rising from a difficult season of loss.

Loss related to my identity as a clergy person; leaving behind beloved community with whom I shared life as a pastor, the divorce proceedings and fighting within my home denomination and her churches, and a new clergy appointment with high hopes that were quickly replaced with challenging dynamics that would eventually tear at the heart of said congregation.

Loss related to my identity and understanding of home as I faced the loss of family members, long-term friends, spiritual mentors, then my mother's sudden death, and my father's death a year later. All these vacancies in my life were accentuated by the bouncing back and forth between the demands of the church, moving in and out of residences for varied reasons, taking care of my aging parents, and the ongoing family dynamics of grown children and a growing number of grandchildren. Even with the care and understanding of my husband, Glenn, and the presence of the home we built on the corner of the family farm, I felt like a homeless orphan, overwhelmed and under impressed.

Losses that brought me face-to-face with who I thought I was and where I found myself. Now older than I can count on my fingers and toes several times over, I look back on my mistakes and count the lessons learned and the lessons I am still learning. And, the voices of the world seemed to crowd out the sacred, still, small but power-filled Voice. An imperfect Child of God, trying my best to serve the congregation I truly loved. However, the love I thought I was giving was "not enough" for some parishioners. I know what God says about His love and mercy and grace and faithfulness to me, yet that roadway was hard to navigate in the persistent fog of heaviness I felt. I was quickly becoming depleted from a physical, emotional, mental, and spiritual perspective.

The practice of gleaning from the fields after the harvest is not practiced as much now. To go back to a harvested field, a field empty of its fruit, and find bits and pieces of life-giving grain still available. To provide for the poor in spirit, offering hope found in pieces.

This is the season of gleaning the empty fields to find the hope that is needed, the grace that is offered, the love that is waiting to be brought to the table.

Jesus tells us Blessed are those who hunger and thirst for righteousness, for they shall be filled. (ref. Matt. 5:6).

Loss makes us desperate. Loss leaves us longing for days when we feel loved. Loss draws us to the one who can satisfy our hunger and thirst for life again (Isa. 55:1-2, John 6:35).

By the end of December 2024, during a heart-wrenching time of prayer, I heard the Holy Spirit tell me to stop, rest, and renew before I went any further. I could no longer pour out until my earthly vessel was repaired and re-filled. And so, with the agreement and blessing of the church leaders, I took a much-needed physical and spiritual renewal leave.

The road from loss to reclaiming life leads into ordinary places and unexpected interruptions. And I was willing to walk that road.

The Letters

OF GRIT AND GRIEF

Dear God,

I'm a bundle of broken pieces, sharp on the edges and barely held together. I want to be at peace, not in pieces. I know you are my Loving Father, and you have a plan. But I'm tired. With each challenging situation I face, I grow wearier. And while the human condition is full of experiences, why is it that emotions seem to take the steering wheel to either run into someone else or drive off the road and hurt oneself? I believe you when you spoke through the Apostle Paul when he said, we know that in all things God works for the good of those who love him, who have been called according to His purpose (ref. Rom. 8:28). In light of those words, what I feel is for my good.

But I've got to be honest, the feeling of being broken is hard. Over the past four years, life seems to have been chipping away at my peace, my joy, my being. I've often gone back to your Word for assurance of how you love me and how I am a masterpiece. I wonder if that's the same intention meant as when others say I'm a piece of work? To emphasize that the Master is still tweaking, changing, and making brush strokes on the canvas of my heart

When I became a wife, life changed. When I became a mom, through the four boys who made it and the two babies who didn't, life changed. When I graduated from college the first and then second time, as I became a teacher, change happened again. And then you called me to be a pastor, life changed. Looking back, those ups and downs and challenges and choices made me who I am today. But who is that?

When the 2020 pandemic shut everything down, predictability seemed to go right out the window. From that point on, my method of gritting my teeth and just pushing on through seemed to operate on an uphill slope.

Then the daughter became the caregiver to the parents. That's when I had to move them out of their home, which was anything but handicapped accessible, and move them in with us. You were gracious and provided a way for them to build their apartment, their own space, for as long as possible. But I couldn't just leave them there.

You remember the night I was an hour away at the parsonage and got the call that Mom saw smoke and, in a panic, called the fire department? When we arrived an hour later, two of our boys had arrived in addition to six emergency units from four different fire companies. The culprit: a light smoke due to the coating on the new heating system coming on for the first time in the new structure – a normal process that quickly dissipates into the air.

Nevertheless, we took them back down to the parsonage with us, complete with overnight bag and five-gallon buckets holding their meds. What a sight we were! It's ok, you can chuckle now – I do. But all I felt was frustration at the beginning of many times they needed me, but my duties to the church I pastored also needed me.

And then you called me to be a pastor – a Methodist pastor at that. I understood that some pastors are called to move from church appointment to church appointment more frequently than others. Itineracy was the name of the game in the Methodist circles at that time, as it had been for centuries before. And, being faithful at each appointment, I tried to stay close to the church flock to tend to them, staying in the parsonage when housing was available.

But then the denomination started tearing at the seams, and a divorce-type separation called disaffiliation entered the scene. In December 2020, I had to end my 18-month term at Nelson's Church due to the church deciding to leave the denomination. At that point, you told me to stay with the mother church even though stormy times were coming. You reminded me of the passage in Acts 27, when the Apostle Paul, aboard a ship in distress, being tossed and twisted, told those aboard to stay with the ship if they wanted to be saved. Nelson was following her call to leave the denomination, and I was called to stay, which meant I had to leave those I'd grown to love at the church – leave just a few days after Christmas in a year that had separated us by isolation and pandemic shutdowns.

But Lord, you blessed me with a place to land for six months until the appointment season started again, and this time I was able to stay near Mom and Dad. While driving back and forth to that three-church appointment, I was still able to take my parents to church on occasion. A fact they loved because they had been unable to go to their home church during the pandemic, and TV church just didn't cut it. Even today,

I get the allure and convenience of online church, but it's difficult to create community from a computer screen interaction.

Then I got the word of my next move. Fifty miles away and the parsonage could not accommodate the needs of my parents. God, you gave me wonderful parents who always supported me and wherever life took me. When I told them, they said, "Go - you have to follow God's call." They weren't always thrilled about my choices or what I did, but they always supported me with love and grace.

"She's always been on wheels and on the go her whole life, so I'm not surprised," Mom would say when she heard about the drive that would be ahead of me.

Feeling like I was abandoning my parents, who didn't drive by this time, I took a deep breath. You called me and you would make away – that's what you always reminded me. So, I arrived at this church, which was twice the size of other churches I had served. I was excited and filled with anticipation as I entered the doors. Then everything I knew and everything I had hoped for flipped upside down. With unique dynamics already in place, my appointment upsetting the full apple cart would be an understatement. My arrival at the church near the ocean not only dumped the applecart but broke it into pieces. And while God used it for his glory and his glory, there was pain and hurt hearts all around. I was determined to keep an eye on you and keep going forward with the mission you had given me. And yet my determined grit sometimes got in the way of receiving Your grace - did then, and truthfully still does today.

It's been said that hurt people hurt people. Not just out of frustration or anger, but mostly out of grief. Grief that change has happened and it wasn't asked for. Grief that a loved one would no longer be around as much. Grief that while I want change in the world, I don't want change in my familiar backyard. Grief that a seed cannot grow until it is broken (ref. John 12:24). At the time, I realized I went into self-protect mode for me and my family as actions, inactions, social media comments, and in doing so, broke myself away from the community. And I did not tell my parents about the accusations, misinterpretations, or even outright lies talked about me and Glenn because I didn't want any more hearts hurt in this season. Bit by bit, I was losing my confidence in who you called me to be. In looking back, I see where you used this time to draw me closer to you.

While time heals wounds – because time gives us space to catch our breath and step back away from the self-focused response – those wounds still leave scars behind. Jesus healed people who were hurting – grieving people. People who liked the leper had been rejected by those with whom they shared the community. People like Zacchaeus, who found himself on the outside looking in. People like the nameless, whom you see even when others don't. I don't know all their brokenness, but you do, and you care.

You heal – not for my comfort but for me to know you more. Your own Son was healed after three days in a tomb– all the marks made by the whips and sharp-edged stones were gone on Resurrection morning. But the marks on his hands and side remained. Scars that tell a story. A true story for a true purpose – and while that purpose includes me, the purpose is not about me. And if my version of three days is now, then so be it.

> *God, I wish I could say that in those days I was singing. But I'm no psalmist, and there is no song, no words, nothing rising up. But in this journal, this letter, this prayer to you, I'm calling out. I'm here, Lord, waiting. Waiting for You. Waiting on You. Waiting for You. Lord, hear my prayer. Love your child.*

Love letter from God

My child, you are right where you need to be. When you come with empty hands, you are ready for me to give you the good things I want to give you, the good things you need. My presence, my guiding rod and protecting staff will comfort you. The appetite for more of Me will come for if you seek me, you will find me. If you ask, you will be given. But for now, rest. You are weary, weak, and heavy laden. That was never my plan for you. Rest in me and we will journey through this together.

Blessing after the Meal

May the One who strengthens you in weakness walk with you through every storm and whisper peace to your striving soul.

LOSS OF APPETITE

Dear God,

Today, I drove past Rudy's Family Restaurant and remembered Mom. Not because she liked their Greek salads, because you know she did. But because Mom was always where home base was and when that changed, my world changed. Remember that day? Those events are seared into my mind.

"What do you want?"

What do I want?! I want to ask questions only she would know. I want to see that loving look she gave my Dad. I want to make up for so many missed opportunities. I want my Mom back.

"Mom, the waitress needs to know what your order is."

My son's voice reminded me where I was, as I looked up into the face of a very patient face of teenage waitress. With compassion on her face, she gently asked," Have you decided?"

Faces ranged from lovingly understanding to frustrated over over-hungry looked at me, because without my response, they were not getting dinner.

"How about liver and onions, with applesauce on the side?"

"Ewww, Grandma, that's gross." Out of the corner of my eye, I caught the slight curve of a smile on my Dad's face. Because if you know, you know.

19

In my childhood, my mother would bribe me by putting applesauce on foods I didn't like so I would eat the required "two bite minimum." Over time, applesauce became less, and the list of foods I liked became greater. Applesauce and ketchup saved me over the years, but now nothing could cover the spreading emptiness of grief. Yet I tried to hide what I didn't want to deal with, the applesauce of denial and the ketchup of distraction.

Sharing a meal on the night before Mom's Celebration of Life seemed appropriate. To gather around a table together is what we did as a family, all four generations present. And to share that meal at Rudy's Dinner hit an emotional chord for all of us. When the crowd was too big for the dining room, and the weather was iffy, we would all go to Rudy's.

At Rudy's, everyone could find what they wanted or needed. At Rudy's, I didn't have to cook or clean. At Rudy's, all were welcome, and if one found difficulty in paying the bill, a way was always found.

Rudy's was a family restaurant where the proprietor, Rudy, and his wife, Dilek, raised their children. In the early days, a charming photo of their three little girls graced the corner of the back of the menu. Years later, those same girls, now girls-turned-young ladies, help with the family business when off from school, including one of them being our server that day.

Family, at the table. Sharing memories. Laughter and stories filled the air, even once the food arrived. And yet when the food came, I couldn't eat. Dad seemingly felt the same way because he only picked at his BLT on white. Oh, we nodded in agreement on a memory shared, but the person we wanted there wasn't there anymore.

No appetite for the food, and to be honest, my appetite for life had greatly diminished. Don't get me wrong, I am thankful for the family and friends seated at the table, but one was significantly missed.

The next day at the service, I followed Mom's directions to the letter. I was tossing one of the special music selections she had written down five years earlier. Planning was made easier because of her prepared directions on who was to say what, sing what, and even down to the details on the flowers.

If I go before Dad, make sure that there is a red rose placed in his hands stating, With love always, Beggy. And since your Dad won't remember, if I go first, there should be a yellow rose in a similar position, including "With love always, Joe." And that made sense. They had always finished each other's sentences, shared the same taste in sweets like chocolate-covered graham crackers, and held hands whenever they got the chance. A timeless love for over 63 years. And now a love that had been separated by death. My father was feeling so lost, and I couldn't make it better.

I couldn't say "I'm sorry" enough to my Dad because I felt guilty about Mom's last day. Then that reel played again in my mind.

It was a typical Thursday morning. I had overslept, and it was now 6 a.m. I needed to get my shower, get dressed, and get out the door in quick fashion if I was going to be on the road by 7 a.m. Even then, I was going to be delayed by school buses and hurried workers, so the morning commute would likely extend to 80 minutes.

Upon exiting the shower, I noted my phone had a missed message. My parents lived in an attached handicapped accessible apartment we built for them five years earlier. Mom usually was an early riser, but for her to leave a voicemail was unusual at this hour. Upon closer inspection, she had also left a text message. So much for getting down to work early. So, I called her while I finished putting on her shoes.

"Morning, What's up?"

"I need you to come over. I had a really bad night last night."

"Ok. I'll be right there."

Placing my black messenger bag with my work notes in the car first, I walked down the cement ramp we had put in that connected the main house and the apartment. Reaching the apartment doorway, while the glass door was closed, the interior front door was open.

Upon entry, I saw my mom sitting on the couch with my dad, holding hands as they always did.

"Are you ok? Are you in pain?"

"I had a bad night - couldn't sleep. I don't hurt, but I don't feel right."

I couldn't find the blood pressure cuff, but I grabbed the O2 monitor.

The display read 82, and that wasn't good.

"Are you willing to go to the ER? Or do you want an ambulance – I can call them."

"Yes, I'll go to the ER. Your Dad should be ok here until Suzanne gets here." (Suzanne is her sister who was coming for a visit that morning.)

"Ok, let me run next door and get my purse, and I'll pull the car around."

On the way back over to the house, Suzanne called, telling me how Mom had called her and then sharing how some of Mom's symptoms sounded similar to the ones she had before her heart attack. I told her Dad would be home while I took Mom to the hospital, 25 minutes away.

Time was now more of the essence. After pulling the car to their sidewalk, I went in and helped get Mom's shoes on her feet and put on her jacket as the chill of a fall morning was in the air.

"Can you walk to the car?"

"Yes," she said in a tired voice.

With Dad on one side and me on the other, she used her rollator to walk to the car. Once there, we helped her in

"I just want to lie down."

"You can't, Mom, it's not safe. I'll recline the front seat, but I have to hook you in."

Snapping the seat belt in place as Dad kissed her and told he he'd see her later, I made sure Dad safely got back in the house, then quickly got in the driver's seat, headed to the hospital.

Mom's eyes were closed.

"I'll get you there as fast as I can."

The passage of time is often hard to measure. At times, minutes fly by. At other times, a minute seems like an eternity. I don't know how long I was on the road when I heard the sound I'd heard before as a pastor. The sound is when you are standing by the bedside of someone who is actively dying, and all of a sudden, the last breath with its little gurgle is heard.

Three breaths with a little gurgle are what I heard from the seat next to me.

"Mom…"

"Mom, are you ok?"

A shake to her shoulder with my free hand. Nothing.

Now, why I didn't think to call 911 or drive to a fire station, I don't know. I'm guessing a part of me hoped she just passed out. That's what I thought had happened, didn't just happen in the front seat of my car. And by the time I stopped somewhere, waiting would have cost precious time.

When we arrived at the hospital ER entrance, I tried to get her to respond again, but she didn't. So I quickly went inside to the empty ER waiting room and told the nurse

I need help. My mom is in my car and not responding.

I guess the nurse knew from my expression and tone of voice the seriousness of my statement. Within a few minutes, a team of six clad in scrubs came out to the car. Opening the front door, Mom's arm fell out, and as the nurse checked her pulse, her face told the story.

The team put her on the gurney they had brought with them and took her to a trauma room, while one escorted me to the separate private waiting area. I knew all about those rooms. For families to wait to see if their loved one lived or died.

I set out to notify family and close friends of the situation. About 15 minutes later, the doctor came in to tell me she had died and they couldn't get her back. I knew she was dead, but now it was confirmed. And after some tears alone in that room, came the hardest part. Telling my Dad over the phone so he could come see her one last time.

My aunt arrived at their apartment and brought Dad to the hospital. He came in a wheelchair because his gait was unsteady. From the moment I saw him coming down the hall, I could see the reality of the situation hadn't yet sunk in.

"I'm sorry, Daddy. I'm so sorry. I tried."

Hugging him, I moved to his side as we went to see Mom lying on the table. The sheet was pulled up so he couldn't see where the emergency tracheotomy had been performed.

With the realization that his beloved was no longer alive, my father's deep sobs hit my heart so hard.

Tears began to fill my eyes, then I remembered I was still at the restaurant, although now the rest of the family had finished its meal. That is all finished, but my Dad and our food were cold and barely touched. So, with takeout boxes, we packed up what was left and headed home to get some sleep.

The day of her service would have met her approval: granddaughter by marriage Roni fixing her hair, the casket with the praying hands she had picked out years earlier, her wearing the sparkly rose colored pant suit she had worn to my youngest son's wedding just four months earlier, the flowers of many colors on top of the casket because she could never just pick one favorite color, her four grandsons sharing their love for her, and her friends from work, church, and life in attendance as well as some folks from my church. I sang that day but couldn't give a eulogy – instead, I sat by my dad's side as he greeted everyone and smiled at all who came to honor her. At the end of that day, I stood on their porch and saw the most beautiful sunset. Well, that's one day down. A lifetime to go.

God, I realize heartbreak is a part of life. That reality is splashed all over the pages of Your Word. But for a heart to be broken, it first must be willing to love. Your heart broke when you saw you only Son, not only betrayed, not only lied about, not only whipped and beaten and whipped again, but then when made to carry a cross not of his own making, have huge, angry spikes nailed into his hands and feet for weightbearing purposes all so he could painfully die a sinner's death. And what did he do? he forgave the thief next to him. He comforted his mother by making sure she was taken care of. He called out in honesty to you. And you had to let him do it as part of your plan to reconcile us to your perfect and Holy Self.

I saw Mom die and wondered if it was my fault. If I were to blame. But what you have spoken to me is that you know every hair that is or ever was on my head, and that means you knew she would die on that day. That her prayers not to end up in a facility, to go before my Dad in death, to not be in pain or be afraid – you met her there and were with her.

> *God, while that broke my heart, her death just reminded me you know all things. That when I don't have a stomach for what happens, you never do. When I tend to give up, you never do. When I complain and, in my humanness, try to run and hide, you never do. Thank you I am not God and you are. And if I ever think I am, shake me up and tell me to get over it. You alone have the full clear view of the past, the present, the future. I have no need to fear when I put my trust in you.*

Love letter from God

I do know all things in heaven and earth, and love you even more than the sparrows in the sky. You do not need to hide because I know where you are and will continue to call to you like I did with those who sinned in the first Garden. But remember, like happens in a Garden, for that is where my Son prayed for you as he faced death, that is where he rose and conquered death for you. My love for you doesn't die, so live into your questions, for when you seek me, I will be found.

Blessing after the Meal
May the God of all comfort meet you in the ordinary places, filling every empty space with His tender presence.

ON MY OWN

Dear God,

I find myself taking time out for you, and I don't know what to say. My journal pages have stayed blank for weeks, even a couple of months. My quiet time prayers seem impotent and trivial if anything. Speaking of your greatness and your glory is a proclamation I make every day, but getting real and in the dirt about my layers of feelings – that brings me to a level of vulnerability I fear going.

I am very familiar with the ACTS of prayer: start with Adoration, then Confession, then Thanksgiving, and finish up with Supplication or Stuff on my Prayer List. And I say the prayer you taught your disciples to pray. And sometimes, the "God is good, God is great" grace. But then what? I can say all sorts of things, but to get to the heart of the matter, I have to go deep into my heart. And that worries me. I know I am a sinner saved by the grace and forgiveness of Your Amazing love through your Son Jesus Christ. And that Your Holy Spirit is with me. Even when I don't feel your presence or see you clearly, I know you are there. Experience tells me that.

Remember the time after my house burned down in Delmar? My youngest was only a year old, and now he has a baby of his own. Yes, that night was so long ago, but I still can see the charred ruins, and even more so the smell. And not just the smell of burnt wood.

As a school teacher, one always wishes for snow days. But not in the manner it happened that night.

I woke up suddenly. And without even putting my glasses on, I could see the clock wasn't working. The electricity must have gone out. Oh well, I'll start the day earlier than anticipated. So, I walked barefoot out of the bedroom, down the hall, and before turning toward the back of the house where the electric panel was located, I saw a light. A glow.

Wonder if only some of the breakers went out.

I turned toward the living room at the front of the house, where the rocking chair was engulfed in flame. Survival mode kicked in, and – still without glasses – I woke the entire household up.

"The house is on fire. Grab your brother and head for the door."

"But it's snowing. And it's cold."

"I bet there are six inches out here. think we will have a snow day?"

"Go get in the van. Here – put on one of these coats."

Grabbing the coats on the rack by the doorway was easy, but there was no time to linger.

The family van was in the driveway next to the porch where that pajama-clad group of boys exited the burning structure. Grabbing the dog by the leash from under the kitchen table – hiding, I understand – but not the best place at the moment – we headed out into the cold night. Now up to this point, I did not feel the heat, hear the sound, or smell the fire. Yet the moment I stepped out of the front door onto the side porch, all systems came online. My ears were filled with the crackling sound of the fire, while I felt the heat and smelled that acrid smell. Six steps down to the sidewalk, and the windows blew out from the living room we had all just passed through. Oh, and did I mention with no eyeglasses, no shoes, no coat, me clad in my flannel night gown clad me holding onto the dog's leash. Then, when the windows blew out, Duke, the yellow lab/pointer mix, decided he had had enough and bolted hard enough to take me to the ground. My bottom hit that snowy sidewalk, and there was no cushioning effect.

At that point, the boys started getting out of the car as neighbors came onto their porch and called to them. It was probably a good idea since the fire spread to the van within minutes due to the gusting winds of the snowstorm. We all survived with the clothes on our backs.

While making phone calls and arrangements, I found a receipt in my coat pocket reminding me I had gone grocery shopping the previous day. Like a week's worth of groceries were sitting in that refrigerator in my now blackened kitchen. A few days later, I went back to the house to find what I could. However, when I opened the refrigerator door, the stench of rotting food was so bad it made me gag. All that good went to waste because of a faulty electrical socket.

God, you woke my mother-in-law up and prompted her to pray for our family that night – not knowing the details, and she did. To have the faith to trust that you will take care of the details – that's the faith I need now.

While my only brother, two years older, had come up for my mom's funeral, his life and work were hours away in the Deep South. As the power of attorney and his only daughter, Dad now fully relied on me. And at that moment, if he had his way, he would have gone to sleep and woken up with Mom in heaven. But that wasn't God's plan or timing.

Now, instead of two households to manage – the parsonage and the house near the farm – I was Dad's secretary, chauffeur, and daily reminder of Mom. Doctor appointments were in abundance, and now with mom gone, he was no longer willing to go anywhere on the Senior Center bus. What had once helped me balance their care was no longer an option. So, I had to get more creative and was thankful for those who helped in any and all ways. He missed her so much that some days he had no appetite and wouldn't eat anything, or the anything he ate was his beloved sweets. Maybe that is where I got the habit of sweets as comfort food.

Anything to fill the void. Yet, isn't that what You warn us about in Your Word?

Not to let anything become the ultimate. That you alone are worthy to be honored, you alone are worthy to be praised. Your grace is sufficient. And yet in that moment of weakness, my sufficiency was found in sweets like favorite cookies, a spoonful of marshmallow fluff, and a piece of chocolate because it's only one bite. Yeah, I bet that's what Adam and Eve said, and they didn't even know what marshmallow fluff was. Regardless, I paid the price from a health perspective. Add that to the list of things that need attention.

Lord, you know my second year at the church near the ocean was its challenge. Leading a church, a grieving church, is difficult when the pain is in layers: the losses experienced in recent pastoral changes, caught in the cultural pull and tug between political positions, and struggling with decisions on how to respond to the denominational disaffiliation question with its ticking clock. And there I was -still wary of my vulnerability to this relatively new group of people, nursing raw wounds from the previous year, and taking care of my dad, who was now at the house 50 miles away from my work with few available to step in when I was away. Not only did I feel pulled in different directions, but so did the church and so did my Dad.

God, when Paul and Barnabas split ways over differences, you created more opportunities for your gospel message to go out into the world. Then why is it today, having differences seems to be an excuse for the cessation of something rather than the creation of something new?

And then the loneliness factor in the middle of all the cultural chaos. I think it was a line in a 2009 movie (ref. "World's Greatest Dad") that captures the feeling: *I used to think the worst thing in life was to end up all alone. It's not. The worst thing in life is ending up with people who make you feel all* alone.

Dad was alone for many hours in a given day. Glenn was alone, driving in his truck for long days. I often felt alone in a room full of people. And while I sought You and You were ever faithful again and again, I was physically and emotionally wearing out.

By the end of the year 2002, those whom I had been ordained with or served with or even attended meetings with had gone their way, and I had gone mine. A few went in the same direction as me, but it seemed none were close by. Again, I felt on my own in a sort of wilderness. A new start would come from the wilderness, but when you are looking out on the barren landscape, hope seems to hang on by a thread.

Maybe that's why I resonate with Elijah. Not because I'm some great prophet performing miracles in defiance of hundreds of enemies. But because, like Elijah, I reach a tipping point. For him, after that glorious God moment when You showed up large and in charge in the most spectacular way (ref. 1 Kings 18) by giving courage in the face of danger, defeating the false prophets, and sending a drenching rain to a thirsty land. Then the very next chapter, Jezebel says, "Elijah, I'm going to get you," and he runs away. Running away to the wilderness and telling you, take my life and let me die." Kind of dramatic, but it makes the point. And what did you do, let him have a snack and a nap- twice – before calling Him for some one-on-one time on Mt. Horeb.

> *God, I guess this solitude and silence are so important because without it, all I hear is noise and can't hear You. When I simply come, you are simply there. Being on my own isn't a punishment. Being alone isn't a bad thing to fear. You will provide what*

I need when I need it. The entire Bible is a witness to Your love for me. My life is a witness to Your love for me – who else would show such mercy and patience to a mess like me? Open my heart to see that again. Lord, hear my prayers.

Love letter from God

I am the great I AM. That means I am in the silence and I AM the silence. I am in the words of scripture, and I AM the Word. When you abide in me, I abide in you. For nothing can separate you from my love – not death nor life, nor principalities nor powers, nor things present nor things to come, nor height nor depth, nor any other created thing, shall be able to separate us – you from me and from my love which is in my Son Jesus Christ.

Blessing after the Meal

*May the One who never leaves your side speak
a fresh word of life into your weary soul.*

FAST FOOD FAITH

Dear God,

It's been a grab-and-go kind of day. Hurrying to make it to the doctor's appointment on time. Impatient when the traffic light is as green as it can be and the driver in front of me is looking at his phone – not moving. When I realize I left a lab slip at home, and have to turn around and get it. Oh, and I forgot to brush my teeth. Just won't get too close to anyone. And the most important meal of the day is just going to have to be another tea and muffin. Just thinking about all that wears me out.

New Year's Day 2023 fell on a Sunday, and we took Dad down to church. He couldn't stay with us when we stayed at the parsonage because he couldn't manage the steps up to the bedrooms. But when he felt able, we stayed the night at the house, and first thing in the morning would bring him along on the 75-minute ride to the church near the ocean. And on the way, we usually stopped for coffee and food. At times, food was donuts, and other times a breakfast sandwich, but coffee was always involved for Dad and Glenn. True confessions: I don't like coffee, and that is a burden I am glad to bear. For if I drank coffee, so many packets of sugar and cream would be added that it would no longer resemble what my Dad and Glenn had come to love.

Anyway, I was preaching that morning. My dad sat on the front pew, leaning on his rollator. Every once in a while, he would look over to his right, which is usually where my mom sat. They had always sat side by side – in the booth at the diner, in the backseat of the car, in the wheelchairs at the airport, and even when a year earlier they had a dream come true as side by side, hand in hand, they sat in their rollator seats and looked out into the Grand Canyon. That had been four months before Mom died. I wonder if that was crossing his mind on that New Year's Day. At the end of the service, I saw my Dad use his rollator to go up to the altar rail, set it to the side, and kneel on the velvet-covered cushion. And pray.

After a few moments, one of the members of the church prayer team came up to him, and putting their arm around him, prayed with him. Dad was always a private, subdued personality in religious settings. I say that because, as a lifelong fan of theatre, even to the point he earned a high school varsity letter in drama, he was not a person to make much of himself or to draw attention to himself. I never saw him lift his arms in praise or tap his foot during a rocking worship setting. He was moved by the Spirit in the house, but he kept it to himself. The scene at the altar rail was only the second time in my life I had seen my father with bended knee in prayer at church. The first time was forty years earlier when he and our whole family gave our lives to Christ during a lay witness type of revival weekend at the little white country church where I grew up. Now we kids had grown up and moved away, and mom had passed away, and he was here – before His God – having a conversation.

After that prayer moment, Dad changed. In the three months after Mom's death and that New Year

Sunday, he lost his appetite for food and life. Oh, he still had a moment or two when one of his dad jokes emerged:

"Dad, did you eat your donut?"

"I ate all of it but the hole."

Thanksgiving and Christmas had been difficult, and he tried to smile for the grands and great grands, but retired early that evening from the family dinner in my dining room to go to his apartment down the ramp and shut the door. Today, he turned a corner, and his demeanor showed it. He still grieved the loss of his beloved sweetheart, but he was choosing to live each day, one day at a time.

I, on the other hand, stayed busy.

If you are busy, you don't have time to cry

If you are busy, you don't have time for regrets

If you are busy, you won't realize how much you hurt.

I wanted to give Dad something to look forward to more than just going to doctor appointments, so on Fridays I would go into town and pick him up coffee and donuts. Then one day he asked

"Beck, can I ride with you?"

"Sure, where do you want to go?"

"Oh, wherever you want to go."

"Want to go to Dunkin?"

"Sure."

"Want to go to McDonald's?"

"Sure."

And that began weeks of going to both places and getting food from both.

At Dunkin, it was a medium coffee, with cream and Splenda, and an ice cube. Not every order for an ice cube was met, so if the coffee was too hot, I'd pull over after departing the establishment and give Dad an ice cube from my iced tea. (Told you I wasn't a coffee person).

Then we would proceed to McDonald's, where he would get a hashbrown or apple pie for later.

But it wasn't just the coffee and food run; it was me breaking my habit. Dad and I began talking about the songs on the radio. He talked about his days with his mom from the house in Laurel down to the coffee shop where they would spend hours, talking with old friends and making new friends. The coffee shop no longer exists, and most of those who attended those morning breakfast times are gone, but the memories were good ones, and I'm glad he felt he could share them. While some of the stories I heard over and over again, I didn't mind. I became more and more appreciative that I had taken the time that I originally didn't think I could spare.

Losing some of my time off in the mornings was worth it— even on Fridays when the Sunday sermon still needed work. I had been living on a fast-food faith without even realizing it. Gobbling down verses on the run. Listening to scripture lessons via the sync feature in my car. While good at times, the "git and go" approach never works in the long-term building of or strengthening of relationships. And with dad in the car, I had done it again. What I thought was a good thing was not the best thing. It was inviting Dad into my habit.

God, how often do I do that with you? Want you to take my order and give it to me now. I want to call the shots, and you are just waiting for me to my mind so you can do what you have planned. I mean, how many times do I work harder and then say, "Ok, God, where's my blessing – I want it now", not asking your will but my will on my time? Wow, what a brat I've been!

Faith is a process, just like any food that is grown to put on a table or on a tray, or in a bag. If it's a grain for bread or donuts, then that grain takes about 100-200 days to grow and then still has to be harvested, processed, and made into a round-shaped delight. And for some of us not on the vegetarian or vegan mindset, if we are eating a burger, it gives birth to the burger process for about 2 years. And yet, "I want it my way," and I want it now. Wow, what do you have to deal with 1

God, you are so good and so patient with me. Forgive me that I forget it's not about my timing. That, in the big scheme of life, while I like next-day shipping, it will not give me the results that change lives. It will not give me what I need.

Forgive me for allowing myself to become weary and selfish because my timing was not your timing.

And my striving was in vain. I lost sight of you and of the gifts you have given me – like the different beautiful people around me, including my Dad, who too often got my leftover time, not my intentional time.

> *God, is that why you tell us "Where two or more are gathered, you are there"? O, hear my prayer of confession as I seek your merciful and gracious face. I realize taking time to be with you is not just about meeting together with others for church or a prayer meeting. Those are important. But taking time to be with someone on their schedule allows your love to flow through them to me and back again. Being present for someone else is where we hear you more clearly. Forgive me for being so focused on my to-do list, I forgot to be with you and the others you have put into my life on purpose for your greater purpose.*

Love Letter from God

You will always go through seasons as long as there is winter and spring and summer and fall on this earth. But remember the lilies. You do not need to worry, nor do you need to rush about in such a manner that drains your spirit. Life is more than just food and clothing, more than what you do for a living. I take care of the birds of the air and the lilies of the field. And you are of greater value to me than them. Trust me to give you what you need and the time to accomplish it under my will. Then my peace will settle in and feed your weary spirit.

Blessing after the Meal

May you taste and see the goodness of the Lord
In every sigh, every breath, and every divine interruption

BLUE RIBBON SPECIAL

Dear God,

When you inspired the Psalmist to write, "taste and see that the Lord is good," were you thinking of pancakes with maple syrup? I may sound silly, but hear me out. Whether it is pancakes or a savory stew or even freshly-baked bread, all those tastes and aromas draw me physically to the table that is set. How much sweeter and richer and filling is your presence to draw my soul nearer to you? And how much is missed when I don't take the time to recognize the food isn't what I need as much as the relationship of the one at the table.

In those days when Dad and I would do our coffee and donut runs our time together was filled with stories and laughter. But at the end of the drive, we both went back to our respective to-do lists for the day. Thankfully, you taught me, through Dad, that the relationship didn't end there.

On one of my Fridays off, I went over to Dad's apartment as had become a habit.

"Hi Dad, ready to go? Will it be donuts or McDonald's or both today?"

"If you have time, and if you want to, I'd like to go to Rudy's."

"Rudy's in town?"

"Yes, I think they have a Breakfast special I want to try. If that's ok."

"Sure, it's ok."

Rudy's Family Restaurant was located on the dual highway that ran just on the edge of town. While originally operating as a 24/7 establishment in its early days, the COVID pandemic of 2020 changed the operating hours to the standard 7 a.m. to 9 p.m. After all, the town was typical of most small towns with sidewalks that rolled out at sunrise and rolled up after dinner time. Those who went to work early either ate at home or grabbed something from the fast-food option that opened its drive-through window from 5 a.m. to midnight. But to stop and sit down for a meal, even the fast-food option was a 7 a.m. to 9 p.m. offering.

We went to Rudy's that morning and soon joined the ranks of others who wore the honor of being regulars. You know, the regulars who were known and greeted as family. The regulars who didn't have to ask to see a menu because they already knew what they were getting. The regulars who were known by the staff because they took the time to talk with the staff. In the weeks and months that followed, we became Regulars.

On the first Friday morning, we went to Rudy's. Dad's pace was set. He would always order the Blue Ribbon Special. Two pancakes, two strips of bacon, two links of sausage, two eggs, small orange juice, and coffee. Every week like clockwork.

I began our Regular tenure with different breakfast options, but quickly fell in step with the pattern my father had set. So much so that no matter who waited on us, the question would always be, "So will it be two Breakfast Specials today?" And with a nod and a smile, their question was answered, and the experience of the meal began.

With a cup of coffee in Dad's hand, the experience began.

Wisdom and memories flooded that sacred space between bites and sips. What I originally saw as a longer breakfast than originally planned gave birth to a new level of relationship with my Dad. He was not just the father who had read me fairy tale stories during my childhood. He was intent on sharing his real-life lesson stories. Because without his account of family members and memories past, soon there would be no keeper of the legacy stories. Dad knew the preciousness of these moments with me, even if I was still learning. He still sorely missed my Mom and his buddies back in his hometown, but he found new strength and spirit by sharing over the meal. A meal that became a learning experience. A lesson the adult me didn't realize I needed. Dad's stories were often memories dear to him, like:

- Remember when you were a waitress and you would bring home sweet potato biscuits from work? And that fried chicken was so good. Your grandmom (we called her Mimi) used to have a cast iron skillet she never washed because if you washed it all the time, I was unable to hold the seasoning that made the chicken taste so good.
- Next to Mimi and Pop's house on Oak Street, where that store is now in the building that

used to be a laundromat, that is where Great Grandpop's printing shop was located. I have a picture of you at the age of three sitting on the step in your Easter dress and pretty little white hat with flowers. I wonder if I still have that picture?

- When I was working at the DuPont plant, those who smoked got extra break time, so I bought a pack of cigarettes and had them with me so I could get that break too. And the coffee at the plant? Nothing made it taste better - it was so strong it could have walked out of the building, but it was there and you were there, so that's what you drank.
- I remember when you and your mom and Joe (my brother) would come and pick me up after my shift. Remember that black car with the red inside? "It was the only one we had, and your mom had needed it that day – but I came out of the plant, and there you were, standing outside behind the railing of the parking lot. I didn't always see the car, but I saw all of you.

I had always remembered my Dad as a reserved, mostly quiet man, only making the silly jokes, clever puns, and engaging wit with those he knew best. But here, at Rudy's on those Friday mornings, I didn't just see my Dad, I saw the man God created him to be, free of my labels and perceptions. He was just a person enjoying the moment. Not looking at his phone – although in all fairness, he had a flip phone. Not watching the time on his wristwatch. But moment savoring the time as if it were his last meal and he was going to drink in every drop of grace and taste every morsel of goodness that was to be had.

In those breakfast encounters, I saw my Dad, whom my Mom fell in love with – gentle, caring, funny, honest. But I also saw another side. I saw my father, who, after losing the love of his life, wasn't afraid like I was. It was like, after that New Year's Day at the church in prayer, he had decided to live the life now before him.

Oh, there were moments when he would say:

- Your mom would have liked this.
- See that pastry over there, I remember when you were little and we would take you to the bake shop in the mall. Do you remember that? *(I loved elephant ears and icing-covered cinnamon rolls, but this isn't about me.)*
- This would be a good day to sit on the porch swing? (something he always did with mom as weather permitted, but rarely did after her passing)

And after each reference of her name, the hint of a tear was in his eyes. She was still in his heart and always would be, but he had come to a place of acceptance. He had come to the place where he could reclaim his appetite for life again.

God, you coming to our shared table to share a meal made such a difference. But you know that, didn't you? That's why Jesus often went off to share a meal with those who were seeking, with those he had healed, with those who welcomed him.

Like the breakfast on the beach after the Resurrection. When Simon Peter and the others thought the chapter had closed, you treated them to a fish breakfast over a charcoal fire on the beach. That charcoal probably reminded Simon Peter of his repeated betrayal of you just days before. But you welcomed him to come and eat what you had already prepared for him. Nothing formal. Just a simple breakfast together. And then the best food was offered – forgiveness, reconciliation, the healing after the hurt.

> *God, you just amaze me. When I start to think I know you and know your ways, you teach me, stir my heart, and pull away the heaviness to reveal your refreshing grace in unexpected ways. Don't let me dismiss the ordinary or the interruption to my plans. You take the lead, for you teach me so much more than I knew I needed.*

Love letter from God

You are special to me. Not only are you my masterpiece, but I have known you since I knit you together in your mother's womb. I know every hair on your head. I gave you every breath you breathe.

I know what tomorrow will bring for you, so do not worry. You can trust in Me with all of your heart.

Stop leaning on what you think you know and turn my way so I can set your feet on my path that leads to life.

Blessing after the Meal
*May the Spirit take your questioning moments
and always draw you back to the table of grace.*

PEOPLE EVERYWHERE

Dear God,

Why did you create me to be an introvert? And then on top of it, call me to be a teacher – and then a preacher? Do you know how much humility and grace it took my first Sunday in the pulpit, because I had to ask for something to stand on to see the people? I mean, the pulpits built back in the day were not built for someone five feet two. I know, I know - You have a plan and my ways are not your ways. Guess that's a good thing.

In the beginning, you created all of us in your image. So why does it seem so hard at times to see you in others? Your reflection seems obvious in those who exude joy, peace, patience, kindness, and even self-control. Even meeting a stranger is at times easier than meeting a familiar person who just gets under my skin. I guess that's why you call me to be holy as you are holy. And yet sometimes I feel more hole-y like Swiss cheese, missing the mark here and there.

I've got to be honest, many times I feel like I'm falling short. I don't seem to be enough for others, so how can I please you? But then, I remember my righteousness isn't found in me, but found in you.

In those outings I had with Dad, he seemed to make the mark more than miss it. Oh, I know he wasn't perfect by a long shot. None of us is. But he would not only recognize those with whom he crossed paths, but also had no problem starting up a conversation with a stranger.

Like in Rudy's, after our Friday morning breakfast routine, he would usually move among the patrons like a pastor among the congregation after the Sunday morning service. He would stop by neighboring tables during our entrance and exit from the restaurant and let them know they mattered. Whoever they were, regardless of age, gender or skin color, or style of dress. Leaning in and smiling at the little one with her mother. Tipping his hat as a good morning to the older gentleman with a walker. Extending his hand to shake the hand of the farmer sharing a cup of coffee with other farmers, because the farmers were dressed for work and didn't take off their John Deere caps to shoot the breeze.

"How are you doing today?"
"Enjoying the weather?"
"Did you know them?"
"Nope"

And the impromptu conversations were not just at the restaurant. Waiting rooms of doctors' offices and medical centers also became sacred spaces to meet and greet with a smile and a nod. People of all walks of life are interested in my life because of a specific need: the need to heal.

Healing of the body from the seen and unseen effects of a broken world.

Healing of the mind from traumas often buried under self-protective layers of denial, loneliness, and anger.

Healing of the systemic wounds a society inflicts when one is placed into a "less than" status based on biased encounters.

- The sweet child whose face had such an extreme skin condition that a classmate called him a monster
- The child who couldn't hear and was left to figure out the conversation, and in some ways, the ways of the world on their own.
- The single parent has to explain to their child why the cost of involvement in school sports or band, or other activities is just not going to happen because there is no extra money in the already thinly stretched budget.
- The teenager who is frustrated by the church that says it loves, yet those of different skin color are not invited to the pool party.

Too tall, too short, too skinny, too curvy, too quiet, too light, or too dark.

The differently abled, physically challenged, and emotionally challenged.

Rejection in any of these areas creates wounds that go so deep over the years, one might think they are forgotten. I thought the hurts of the past were done and over, but what I realized is that at the core, they affect how I think of myself.

Past labels and words, and actions were getting in my way of my fully loving you and loving all neighbors. Comparing myself to others disrespects the divine and sacred within me. Still, the lure of acceptance tells me the lie that my identity is determined by the world.

My identity is found in You who created me and knew from day one that I would not like Brussels sprouts or analytical geometry... yet still you love me so.

God, no one wants to feel the sting, even the scars of rejection. I guess that is why Your Son showed us that even when rejected, even when facing bullies, even when abandoned – you have a plan. A plan bigger than the moment of hurt or broken hearts. For in the darkness, in the brokenness, is where You create life. Like broken shards of colored glass that are a hurtful mess until brought together in community to create a beautiful retelling of sacred truths.

In the Spring of that year, he was feeling up to going to my church's biannual luncheon for those over the age of 70. I was only able to attend because I was the pastor, and I was the transportation for my father, who was 86. This time, he was the one greeted and welcomed in, and he accepted with a quiet joy. The songs and jokes shared were ones he remembered from years gone by. Nostalgia filled the air and, in that moment, all present were swept up by a lovely meal and a greater sense of belonging. Most in attendance didn't want the time to end, but after two hours, naps were the next line of business for the attendees, as well as those who had worked long hours before the noon time start to make sure everything was just right. Yes, Dad was in his element, and for a moment, grief took a back seat.

Again, something about coming to the table joined people together. Maybe that's why my lifelong experience in Methodist churches always included food that was not just a physical nutrient but a spiritual conduit, giving reason and purpose.

Like when David invited crippled Mephibosheth to eat at his table, not for what the child of his slain friend could do for him, but how they together could honor God and those who had been in their lives. Or like when short Zacchaeus was hanging out in the branches of the sycamore tree and Jesus saw him and called to him, "let's go to your house and break bread together." Or when you showed the power of breaking bread in the upper room and then to two travelers headed to Emmaus. Breaking bread shares food, but a willingness to be broken is a willingness to share life authentically for the good of all.

> *God, you see me...not from afar but intimately. Not just passing through with a nod of acknowledgement, but stopping and staying and seeing. Forgive me when I don't see but don't see, when I hear but don't listen, when I speak but don't communicate care and concern. Your miracles are not about me and the moment, but about you, the Almighty, who has the power and desire to draw me to yourself. Give me your eyes, your ears, your words, your heart for your glory. Hear my prayer, O Lord.*

Love letters from God

Beloved, My love covers a multitude of sins – selfishness, self-righteousness, indifference, prejudice, apathy, hatred – and my disciples are constantly learning to love like Me. Loving others won't always be easy, but by my Spirit, it will always be possible as you seek to be holy as I am holy. When you love your neighbor, you learn to love Me more. And when you love me more, you will see yourself in the light of my love and grace. That may seem overwhelming now, but in time, my love will transform you into the child of mine you were always meant to be.

Blessing after the Meal

May every encounter today lead to a closer walk and a clear view of Jesus and the life He offers.

TOUCH OF PAINT

Father God,

I put on my black comfy yoga pants today because I wasn't going anywhere special, and that's when I saw it: a random stroke of white acrylic paint that didn't make it to the canvas. I still have Dad's paints and brushes sitting in the corner of my office. I haven't been able to crack open the lids or make that first bold stroke onto a pristine white canvas, full of promise. I can't go there yet, but maybe one day.

You made my Dad with an artistic eye. He found the beauty of people and places around him through the lens of his camera, through the pen and charcoal drawings he would call scribbles, and through his painting. On the wall of my home, the first thing I hung on the wall was his still life oil painting he created when I was one year old.

A bowl and pitcher, with fruit, on a table.

Kind of ironic that his drawing of a set table is what reminds me of you.

Dad was 27 years old when he painted that image. And he didn't seriously get back into painting again until after his retirement, about thirty years later. Oh, he was the go-to person for creative decorations at church or in my mother's elementary classroom, even the occasional sketch for the DuPont plants newsletter. But after retirement, he took classes at the community college because he thought "no one wants to look at that mess," speaking of his artistic endeavors. I found award ribbons from art shows I had forgotten about; unfortunately, at that time in my life, I was married and raising children in a neighboring town and didn't always think to ask about what he was doing. Mom probably told me because she was always very proud and encouraging of my art. But when she retired a decade later, knowing their house on Oak Street was small, he chose to let her have the space for her Bible studies. Paints, easels, and canvases take up room at the kitchen table and can't always be quickly put away for the next meal. So, another couple of decades passed before he picked up the painting bug again in 2023.

In the Spring of 2023, Dad had a biopsy done on a suspicious sore, and it was cancer. During the biopsy, the doctor could not get all of the cancer cells. Since radiation and chemo were not an option, Dad had two choices: radical aggressive surgery or leave it alone and hope it didn't come back for a long time. At the age of 86, he opted for the second choice, and I supported him. After that surgery, he physically felt better and asked one day to take a ride to the hobby store.

Lord, you know that in addition to meals and doctors' appointments, we also did grocery store and department store runs. Picking up a few items here and there. But when Dad wanted to go to the hobby store, "just to look at the painting supplies", I knew hope and joy were once again rising to the surface. Yes, you know he bought that 70-inch big TV for himself at Christmas because "your mom never wanted a big TV in the house, but I'd like one," and I, now, got to tell him, "If you want it, get it. "And like a tickled kid, he put up that huge TV whose feet were within a pencil width of the edge of the long console where it sat. "I guess that's big enough", he said with a grin. Joy upon joy was lighting up his apartment.

That spring, we made several trips to the hobby store and came home with canvases, frames for the canvases, paints, brushes, more paints, and then other painting stuff I never quite understood. Modeling paste. T-squares. Sponges. Sprays and such. He was in his glory, and that made me smile.

One day, I came home from a morning at Church and he had painted three small canvases. At that point, I started posting pictures of the works of art on social media. Comments came from everywhere and, while I forwarded them to him on his Facebook account, I think he reveled in my reading them to him after our evening dinner: How creative. I want one of those if possible. He's so good. Love the detail. Beautiful. And his response?

"Did they say that?"

"They are just being nice."

"Those (paintings) aren't that good."

Our Friday breakfast that Spring into Summer often led to his encouragement for me to find my inner artist. Wearing his large shirt like I did when I was little, I began picking up the brush and following its lead. Learning about base coats and Payne's grey. Learning about the forgiving nature of acrylics, that if you mess up, it can be cleaned up immediately. (Oils are much less forgiving, he would say.) Granted, every once in a while, a stray dot of paint would choose its direction of expression like the white on my black yoga pants and specks of red on his blue jacket hanging on the back of the kitchen chair. And when we finished, we'd wash the brushes and the white dinner plate turned artist's palette and call it a night.

I was painting a few decent works on the 8 X 10 canvases, painting that was elementary but promising. And Dad, after looking at it, said, "May I?" and then, with a touch of a paint pen or brush of a Q-Tip, took my elementary effort to the next level. And God, that so reminds me of you.

How, when I bring my hurts and dirty laundry of my life to you, you clean them up with mercy and grace, and forgiveness? And when I don't, they seem to set in as a stain that I seek to hide.

How when I bring my humble offering of a sermon or a painting or any effort, you say, "May I?" and then with your Holy Spirit breath and touch, make something more beautiful than I could imagine.

How, when I say, "I can't," Your patience and kindness take me one slow step at a time if that's what's needed.

I believe many pastors are wounded healers. Not perfect in their own right, but taking those broken moments of life and using them to show the grieving family, share a meal with the homeless or poor in spirit, to faithfully share your goodness that continues to draw us back to you over and over again. That's the hope I rely on. That's the truth your children like Henri Nouwen and spiritual directors and shepherds of the faith remind all of us of. That is the reason you became incarnate, God in Flesh, born in a grungy stable on an ordinary farm. That is what the good news looks like in the light of an imperfect world.

I guess like the Apostle Paul, you sometimes take a bright light/ spiritual 2x4 to get my attention. Like Joseph, you put me into my current circumstances and culture, encouraging me to rest in your love and power, even when I can't see the good you might be doing. Like Timothy, you give me a heart for you and a spirit willing to learn from those who have gone before me. To fan the flame of the gift you have put in me – it's in there, you said it is. And I believe you.

> *Father God, help me find the beauty in today and in this moment. To live my todays without rushing through them. To affirm that you are sovereign and have placed me in the family, in the church, in the position, in the circumstances of today to learn and heal, and grow.*

Love letter from God

My dear child, I created you to be a blessing to others. To take the grace and gifts I've given to you, and not bury them in the sand for yourself, but to use them for my glory and honor. Never forget that, yes, you are a mere mortal, but I notice you, I care for you, and I will put a new song in your heart.

For I am your God who is Lord.

Blessing after the Meal
*May the Master Artist guide your hand and heart,
turning your broken strokes into a masterpiece of grace.*

LAST BIRTHDAY

Father God,

Today is a beautiful day. You have outdone yourself. The sun is making everything look brilliant, from the blue cloudless skies to the vibrant green grassy field where the stunning hot pink petunias are showing off your artistic flair. The birds are singing out praise to you. Today looks and sounds like all creation is decked out and ready for a party.

It was June 2023, and Dad's 87th birthday party. With Mom's passing just 8 months earlier, the family wanted to create opportunities to celebrate life together, and so a party was planned. Glenn and I secured space for 20 in the back room at Rudy's. Grands and their significant others, nieces, nephews, cousins, and great-grandchildren with their own table filled the room. Dad was presented with his pair of fire engine red birthday sunglasses, which he promptly placed over his own prescription eyeglasses.

The brunch was a chaotic bliss. Laughter abounded as everyone seemed to have something to say. And while the occasional quieting of a child getting too rambunctious – after all, other people were present in the adjoining restaurant – the mood was light and cheerful. Then his cake was brought out. Thankfully, his granddaughter in love, Roni, only put two candles on it: the numbers 8 and 7. I remember when he turned 70, I had put the prerequisite number of little candles on his cake. Needless to say, other than the heat those little burning candles emitted, a layer of wax had to be scraped from the top of that cake before we could eat it. Thankfully, today, that would not be the issue.

Dad, in his theatrical voice, told the crowd:

"Four score and seven years ago…no, seriously, I thank you all, from the bottom of my heart, I appreciate it. Love you all…

The gathering of all ages echoed back; we love you, too. Of course, my oldest son added in, "Thank you for being born … without you, there wouldn't be a me."

Then, in two attempts to blow out the few candles, he leans toward me and says, "I had to get my breath." Still, he was smiling. A group photo was taken to commemorate the day, and then all began gathering their things and children to depart for departing into the rest of the day. But not before dad stood up, still with his party glasses on, and with his hand in the air, "I wish you all a fond farewell." We laughed, and that moment went into the special memory category of our hearts.

God, it's interesting how every season you create has pros and cons. I mean, like with the land. Winter is a time of rest, but you are working beneath the surface in ways we don't yet see. In the spring, as flowers are breaking through to show off their blooms, bugs and slithery things are also making themselves known. In the summer, as the blossoms turn to fruit, the weather takes on a mind of its own with searing, uncomfortable heat, with the clouds withholding needed rains. Then comes the fall, with full fields and beautiful leaves, but in a matter of days, the days grow colder, and what once looked beautiful leaves a bare reminder.

In this season, while still grieving the loss of Mom, Dad and I have found a new level to our relationship. And I feel like I am entering into my relationship with You, Father God. I still feel pulled between my duties as a pastor and my duties as a daughter, and I'm trying the best I can to give each my full attention, love, and care. Maybe I'm falling short and not giving myself as much grace as you are giving me. Trying to do my best is wearing me down.

Like when I faced the hurt, grief, and loss after divorce, the new challenge and joy of being a single mom was not a new I looked forward to, and yet it made me more dependent on you. Not that one person can ever adequately serve as mom and dad both – that was never your plan. But even in that brokenness, you made a way, and over and over again showed your provision. Whispered hope. Spoke of truth. Gave me strength when I was weak.

While Dad's cancer diagnosis was a wait-and-see approach at the moment, he seems to know something I don't. I have to leave that with you. The situation and outcome are out of my hands.

I know, in this season, you are doing that again. Just like the smell of rain in the air, I can sense that a soaking, cleaning, living water is ready to fall. Holy Spirit, please take these ramblings like you take my inner groaning and lift them as an offering of my trust in You.

Father God, thank you for being sovereign over all things. And for loving me. Thank you for those moments when your beauty shines through a dark day. Thank you for reminding me that you do hear my prayers. That is when I remember, I think about what you have done. I still thirst and hunger for more of you. Help me move from trying hard to trusting more. I want to finish this race of life well, for your glory, not mine.

Love letter from God

Beloved, I created the world one day at a time. You were not meant to take on more than one day at a time. I go before you because I know the plans I have for you, to prosper you and give you a future. I go behind you to remind you that hope in me will be your steadfast anchor. I am beside you to keep you on the straight and narrow path that leads to life eternal. I am with you always, in every season and to the end of the age.

Blessing after the Meal

May the joy of remembered love and the strength of gathered community carry you through the changing seasons.

TIME OF HARVEST

Lord God,

Why is it so hard when seasons change? I will be honest, when I get my warm weather clothes out, I grumble when a chilly morning makes me put on a jacket. And when I wear the jacket and the heat of the day increases, I grumble because I now have to take off the jacket and carry it around lest I accidentally leave it somewhere it doesn't belong.

I guess everyone has their favorite season of time. Whether fall or spring. Whether childhood or adulthood. Whether the time of beginnings or endings. I know change is necessary, and I can feel a change happening in me. The end of one season and the beginning of another, with more unknowns than perhaps I am comfortable with.

That uncomfortable change of season was apparent in the Fall of 2023.

The cancer was creeping back into Dad's life. He was falling more physically, and his body seemed to cooperate less. And the doctor's response was, "That's to be expected." Even his teeth weren't cooperating. The time for Dad to leave the field where you had planted him was coming to an end.

One of our last Friday morning breakfasts at Rudy's bore witness to his decline. Ordering our regular Blue Ribbon Special, he asked for the eggs and pancakes only. As you know, his front tooth had broken in half earlier that week, and full repair would require waiting a few weeks for a dental bridge. I mirrored his order because I had just had a dental implant performed and was also waiting for the dental crown. We were a pair – each with teeth issues that limited our ability to chew. And still we were able to laugh about it and enjoy the meal. But as I rose to pay the check, I saw out of the corner of my eye that Dad was trying to get up on his own, and not being successful. Quickly, one of Rudy's employees came over and helped to steady him as he slowly got to his feet. That would be one of the last times we shared breakfast at Rudy's. Those life lessons would now be taught from a different space.

In October and November, after two hospital stays and a return of the cancer, he was given a choice once more: go into assisted living where he needed someone with him at all times, or go home and enter hospice care. Again, he chooses the latter. Once home, he started the last two paintings but never finished them – he resigned himself to what was coming and said, "I just can't do it anymore".

At first, hospice came once a week for Dad was still somewhat independent and moving around, albeit slowly. I had been balancing work and home time, as I had been the only pastor for the past year at the large church by the ocean, and was struggling to keep up with demands and duties. However, the time was quickly coming when I couldn't leave him at home by himself anymore.

Thanksgiving came, and our growing family came to the house as usual. But Dad did not feel like walking up the ramp anymore to share a meal. We made him a plate with the seasonal fare, and sat with him for a spell, but he didn't want anything. Only to sleep.

By this time, with his permission, we had put a camera in his living room and bedroom to monitor his movements. One early morning, as Glenn left for his daily truck run in the pre-dawn hours, I checked the bedroom monitor. I saw that his walker was not at the end of the bed, so I figured he was in the bathroom just beyond the camera's view. I gave him a few minutes and checked again. I saw the walker was returned to its position at the end of the bed, but didn't see Dad in the bed. I looked closer and saw what looked like an arm trying to reach up from the floor. He had fallen again. Thankfully, Glenn was still in the neighborhood and able to come back by the house because I couldn't lift Dad from the floor. After getting him to bed, I called 911, and when the EMTs came, I explained he had fallen. They checked him over and determined no bones seemed to have been broken. When asked if he wanted to go to the hospital, Dad said, No. Just call Kim. (Kim was the hospice nurse.)

The time had come that I had to make a choice: take family leave from work to take care of my dad, or. Knowing his finances, insurance, and wishes, there were no other options. While preparing to ask the church for leave, I got a mild case of COVID-19. Guess you took care of that situation for me.

Over the previous months, Dad had semi-regular ZOOM video calls with his only surviving sibling, his youngest brother Steve. I would set the time and date with Uncle Steve, and then help Dad connect on his iPad. The calls usually lasted about an hour, and two exchanged memories and family information, even photography and movie trivia. However, that last ZOOM call was the shortest and the hardest. In his weak and stuttered speech, his words affirmed my suspicions that the end was nearing.

"I feel pretty good, but my speech is terrible. I feel pretty good. But I believe I'm not going to feel like I used to …you know. "

"This is what it is. People have been nice to me…. but you can't get off the bed at the hospital, not on my own… I'd fall …I'm not steady…if I fall, it's going to be bad…"

"Love you …you've been a good brother…"

That call was the last time he talked to his youngest and remaining brother.

With Christmas nearing, I read him the O Henry story, "The Gift of the Magi", you know the one, where each sacrificed what was dear in order to give the other a personal and special gift. You know, during our breakfast conversations, Dad told me he had bought that book for his mother – my Mimi – as a Christmas present years ago, but she died 6 days before he could give it to her. Among other books, I pulled out my old, well-worn poetry book and read the poem titled "Lullaby Town ". The poem was the same that he had read to me as a child to help me fall to sleep. And yes, I did sing to him – he and Mom's song (Til There Was You), "You are my Sunshine", and various hymns. His soft smile let me know he heard me, even when his eyes were closed. For his musical favorites like the blues, Big Band, and Sinatra, I turned on the satellite music channel.

Then the last fall occurred as he tried so hard to get up to the bathroom independently; after that, the walker was folded and put into the closet – there would be no more independent attempts. At that point, hospice encouraged me to let family know that if they want to come visit, they may want to do so sooner rather than later.

On the first Saturday of December, his bedroom in that little apartment looked like a king holding court. My brother Joe drove up from his home in Georgia and stayed nearby so he could stop in and see Dad. All of my sons and daughters in love came in those last days. Great-grandparents came in a gave him hugs, handing him the construction paper cards they had made him to "get well soon". Some sat and talked as long as Dad could sit up and engage. Others watched football games on TV, and still others showed him photos on their phone of points of interest in their lives. Even Zuko, a rambunctious boxer fur baby, put his head up on Dad's bedside, with eyes locking as some unspoken message was shared.

The second full week of December, Dad's appetite significantly dropped. Each day, he only wanted one thing with his few sips of coffee. So, he carefully chose a favorite food he wanted to taste one more time. While the request for General Tso chicken worried hospice because of the nausea that often now plagued him, all other requests were granted as best as possible: a piece of coconut candy, fried bologna, a Boston cream donut, and of course, Rudy's Breakfast Special. In truth, you can't get the breakfast special at 5 p.m., and I told him this, so he chose just to try a couple of pancakes. So, I went into town and got two golden plate-sized pancakes for him. And with butter and syrup, he ate every bite. And that became his final meal.

That week also happened to be birthday week in our family. Dad knew this and each day asked what day it was. Monday was my oldest son, Randy's birthday. Wednesday was Glenn's birthday. Thursday was my third son, Patrick's birthday. By Friday, the December birthdays were done. And Friday night, the worst began.

"Blessed are those who die in the Lord, for they will rest from their hard work, for their good deeds follow them..."

I've heard that verse at many funerals, but really, it's the rest of that passage that offers hope in times like this... and we praise God for the victory and the joy he brings amid a world of sin and death. So, we pray all glory be to your name as the one who has conquered death (ref. Rev. 14:13)

I am so thankful for the hospice chaplain who visited my dad a couple of times. I don't know what the two of them talked about, but I felt like I needed to be his daughter in this time of transitioning from life to death in this world. On the last visit, the chaplain, knowing I was a pastor, said, "Your father loves Jesus and he's ready." While that gave me a sense of comfort, the reality of losing my other parent felt like my heart went into my throat, like when you ride one of those carnival games with the sudden drop from several feet up. You know it's coming, and yet your body still reacts in sync with your emotions.

Mom's death had come suddenly and quietly, just a few breaths, and that was that. With Dad, we saw it coming and tried to make the most of every minute. I thought, what a shame I didn't make the most of every minute with both of them before death was even a thought?

Friday night was bad with pain and restlessness. Roni had been helping at night, but even Friday night into Saturday morning wore her out. On Saturday morning, hospice was called, and the nurses came. At one point that Saturday, Dad – who was usually quiet and yielding – was fighting with a strength that came out of nowhere to the point both trained professionals had difficulty settling him. Roni and I both saw the scene and had to leave the apartment before he saw us ugly-cry our hearts out.

By evening, he was settled and hospice left, and I was now on duty for the duration. I made up all his syringes of morphine in advance, like I was taught. To say I slept that night would have been an exaggeration. My eyes closed while rotating between lying my body on the sofa in the living room and sitting in a chair at his bedside, laying my head next to this man I had known all my life.

On Sunday morning, Glenn came over to the apartment, coffee in hand. We turned on the TV and the Sunday morning live stream of my church on my iPhone, but paid little attention to either. Dad's coloring was changing as his breath slowed. Slowed. Slowed. Stopped.

He had gotten his wish. He died in his home. He died in his sleep, with a little help from the medicine. He was at peace. And now he was with Mom and you.

Even writing this now, after all the time that has passed, my heart hurts. Even telling you again – and you know everything that was and is and is to come – breaks through on my brave face, and the salty tears once more trickle down my face. Still, I am happy for Dad, but I miss him so.

> *Abba, did your Son cry when he got to his friend Lazarus' tomb, after he had been dead for four days? Scripture said he wept – maybe that's because no single word can capture the deep, heartbreaking howl and sniffling that goes with the loss of one you love. And you even knew Lazarus would be revived that day to live some more years before he died once and for all on this earth. Hope in you, that's what I cling to. Parents, loved ones, pass away from the earth, in varied circumstances, and affect me in differing ways. But at the end of the day, I come back to you – with my heart in my hands like a kid with a broken toy. And you tell me to leave it with you, you will take care of me. Ok, here it is…all of me to all of you.*

Love letter from God

Dear one, cry your heart out to me. Don't keep it in. I hear your cries, for when your heart is overwhelmed, I am the Rock to which you can run. The Rock that will be your firm foundation for all time. I am your anchor in the storms of life. For I AM the one who was and is and evermore will be. My Son died for you, so you will be with me forevermore, for all eternity. And eternity is now. Take care, I tell you all these things so you may have peace. Yes, trouble comes in this world, but take heart, I have overcome the world.

Blessing after the Meal

May every tear shed find its way into God's hands
as His peace holds you and keeps you in every season of
your life.

WAITING

Lord God,

I am feeling free, but still weary. Does that make sense? I mean, sharing all of these feelings and memories, and even current circumstances, lifts a burden. But when the burden is carried over a long time, even lifting it leaves a weak feeling. Or maybe it's just me. I feel that to some extent now, but even more profoundly in the winter, Dad died.

I had been off work, on leave for a couple of weeks, before he passed to care for him. The long-awaited new associate pastor started on December 1 that winter, which helped me give Dad all my focus. After Dad. I went back to work the next day, as Christmas was quickly approaching in eight days. I had planned to and did preach all of the Christmas Eve services as planned. But called a dear clergy colleague to give the message at Dad's celebration of Life, the Friday before the new year.

At Mom's service, I had been there for Dad. And now Glenn was there for me. I couldn't sing this time if I tried, but I did share the eulogy. And each son, oldest to youngest, shared their love and respect and memories for their beloved grandfather. My Dad, knowing death was coming, had picked out scriptures and songs he wanted, and told us things he didn't want. So as the slide show of a lifetime of memories was shown, the Dad's chosen song, "Goodbye," was played. And the ugly cry started to emerge from within me again. I pulled it together because we still had the graveside service to go to. To this day, I hear that song and get misty.

That weekend, the New Year began, and work resumed. It's interesting how when we don't want to think about something, we keep busy. I wonder if that is why, at Mary and Martha's house, Jesus told Martha to chill.

Martha was distracted by all the preparations and complained that her sister had left her to do the work by herself, and you told her, probably in your calmest, kindest voice... Martha, Martha, you are worried and upset about many things, but few things are needed, actually only one. Mary had chosen what was better.... Sitting at my feet and listening to my teaching, and I'm not going to take that from her (ref. Luke 10:38-42)

Lord, you even tell me to 'be still" (guess that's the biblical version of chill). And while my prayer and devotion time continued, and I faithfully followed your lead in the direction and the care of the church, but found myself avoiding the quiet times of silence and solitude. Because when it was quiet, the layers of memories of what had been started unfolding, and the burdens and heartbreak threatened to take my attention from just moving forward. That stubborn grit and self-reliance are trying to push through again. Thankfully, you broke through at times, but I didn't make it easy.

The busier I was, the less I had to think about the fact that I was now an orphan with no parents, that while Glenn and the boys were there, I had hosted family dinners for my parents. Now there were fewer family dinners, for since they weren't there anymore, I didn't put the effort out as much effort as before. What was my issue – didn't I just learn about how precious the present moments were?

I tried to give all I could to God, to church, and home, but felt like I was running out of steam.

I was trying instead of trusting.

I was striving instead of surrendering.

I was anxious instead of simply abiding.

Waiting is hard. But you know that, don't you? How long did you wait for me to come to my senses? When you gave Joseph that vision of saving nations, yet your timing was 40 years for that vision to come to fruition. Abraham, Moses, Jacob, and even your own Son. You waited 30 years to permit him to go ahead with his earthly ministry, and then not until after he spent 40 days in the wilderness. Waiting is hard because it is reliant on your time and your way, not mine. Guess that is one way to shine a light on my selfish nature of wanting to follow you, but in my time. You always have the better plan, if I just wait on you and in you. Abiding in your terms, which have withstood the test of...eternity.

Like the Apostle Paul, I would do what I don't want to and don't do what I want. (ref. romans 7:15). Over the years, I had written regularly in my journal, and yet few pages bore any honest contemplation, if any writing at all, in the past couple of years. That was when I knew I needed to take time with you, and not just an afternoon. And not just a weekend. I had more spiritual care needed than that.

So here I am, on a spiritual renewal leave, taking time away from my way and falling into your way. Got to be honest, the first couple of weeks, I should have bought stock in Kleenex. Moments I saw as interruptions to my plans to be with you ended up being divine interruptions. And not the sweet, glowing kind of interruptions. The kind that elicited brutal honesty and ripped away layers I had been trying to cling to like a favorite tattered childhood book. Once those worn, tattered pages, I had been holding onto, fell aside, I began to truly abide in you, not my past. Sporadic journal entries of the past gave way to words starting to flow again. Sometimes so fast I don't think even I can read them now, they look like scribbles in some places, but maybe that is my prayer language via pen.

God, thank you for making the pilgrimage to Greece possible. I needed to get out of my own space and walk where you walked with the apostles as you sent them out into the world. Your love, your call on my life, as your child and as a pastor, were reaffirmed time and time again, from touching the rushing stream of Lydia's baptism to communion under the tree in Corinth. Once again, I surrender all and wait on you for the next steps.

Waiting well is worship in and of itself. And in my waiting, I have learned to worship you more. Waiting is the opportunity to grow and deepen my faith. Like the prodigal who had to come to his senses before going home, only to find his father not only looking out on the horizon for him but then running to meet him and throwing a party for him

That is the kind of mercy I needed to acknowledge, that is the grace I needed to accept for myself. And while I'll never know why you called me to be a pastor – you know my impatient nature, my ADHD, my wanting to plan and plan some more, my weaknesses – I believe you called me to show off what you can do and who you are. Even pastors sometimes struggle to accept that grace for themselves. Now is the season to, instead of trying harder, make the time to trust deeper. And that takes courage.

I guess it's not surprising that the last words of wilderness guides Moses and Joshua were for the people to "not fear and be courageous".

Even more so, Jesus tells me to have courage time and time again.

> You are forgiven.
>
> Don't stay in the boat; take courage and step out.
>
> In the storm, take courage, for I am with you.
>
> Do not fear, you are more valuable to me than many sparrows.

And on Pentecost, when the church was born and the Body of Christ sent on a mission, all were filled with the Holy Spirit and told to speak the Word of God with Boldness.

Lord, thank you for waiting with me and for me. You have given me a vision for your church in this time of waiting, and you have given me a revelation of yourself. But more so, you have revealed yourself in my dark and lonely times. You have revealed yourself in my heartbreak and hurting times. You have revealed yourself in my times of loss and brokenness. I just had to look for you. In my memories. In my moments. In my yesterday, today, and tomorrows. Forgive me when my faith was weak and my spiritual eyes were closed. My hunger and thirst for you are not due to a lack of appetite but a renewed desire to taste and see your goodness in the land of the living. Praise be to you, in all ways.

Love letter from God

Loved one, what joy it brings me when the scales fall off your eyes and your hands open to receive what I want so much to give you. I am yours, and you are mine. I have called you by name on purpose for purpose. And I will never leave you or forsake you. Trust me. Believe in me. Find your life in me. Every day. And rejoice in me always, again I say rejoice.

Blessing after the Meal
May the God who holds time in His hands fill your waiting with sustaining grace until the fullness of His promise unfolds.

LANDSCAPE OF HOPE

God's best work begins in the dark.

In the darkness and void, God created the heavens and earth, the land and sea, the garden and me.

In the darkness of the ground, the broken seed transforms and pushes upward toward the heavens, losing itself in order to be transformed into fruit.

The same God who created in the dark brings light as an interruption to our darkness to get our attention at the next new thing emerging.

More predictable than taxes, seasons in life will come and go. And for every loss, space is made for something new to emerge.

The end of a relationship may mark the beginning of healing. The death of a dream may be the seed of a deeper calling. When comfort ends, trust is born. When feeling weary, abandoned, orphaned, or alone, the self no longer satisfies.

My identity as a clergyperson, as a child of my parents, even as a child of God, was flawed. The same God who called light to interrupt my darkness is the same God who calls me by name. Resurrection is not a one-time miracle that God did back in the day.

Likewise, Resurrection didn't end at the empty garden tomb, it began there to point us home. See, my idea of home was incomplete. Home is not just – to coin a phrase – where everyone knows your name and is glad you came. Home is not just a physical location or emotional leaning. To lose those ideas of home is a blessing, for the loss of an earthly view of home makes room. The early church understood the concept when they celebrated funerals as home goings. As long as the earthly view of home is our only touchpoint for safety and security, then we miss the eternal stability and glorious fulfillment of Jesus' promise of an eternal home.

Loss of my past involved letting go of memories that held sway over my being way too long. To acknowledge the past recognizes it as a component of what was necessary for the growth of what will be. Just as the sacraments of the Lord's Supper and Baptism were not just memories, but signs to lead us from brokenness and death to life in abundance and everlasting.

Like the broken bread at the table with Jesus and the travelers to Emmaus.

Like the dying to self and rising anew with Christ in baptism.

Easter Sunday 2025 was the first Easter Sunday in 20 years I had not led a congregation in celebrating the Resurrection of Jesus, and what a blessing I received. I sat with family and sang the songs of the ages, echoing the truth "He Risen from the Dead and He is Lord." I joined voices with children and adults alike as we prayed together. I heard from a preacher other than myself about the Resurrection power and the victory over death that wasn't just then, but is available now.

And then there was one more piece that brought personal healing.

My youngest grandson was brought by my son and daughter-in-love to the front of that little white country church by the side of the road. The child was born one week shy of a year to the day that marked my father's death. As Reverend Grandma, I was invited to take part in the baptism of this little one. I walked forward and heard a voice speak to my heart, "I am here." For a moment, I thought it was Jesus, for I know Jesus is present in the sacrament. But the voice wasn't singular. The affirmation came from the great cloud of witnesses who were cheering us on from the other side of eternity: the child's great-grandparents, who included my Mom and Dad. And I was overwhelmed in the moment by the recognition that, while I was far from being alone, I was loved.

The same voice that spoke over the deep. The same voice that spoke to broken pieces and dry bones. The same voice that spoke into lifeless clay. That same voice that spoke into my hurt and loss was speaking new life into me that day.

Months and weeks after I stopped listing my losses, the healing was taking hold. A healing found by being present in relationship with God, and showing up to be in relationship with others. Showing up with courage, even in pain, even in grief, even when I didn't think I could.

I learned over and over again the truth I knew but at times forgot to apply in my own life: God is with you and for you and meets you where you are. Even if it's in a church pew, on your bedroom floor, sitting in traffic, or even at a table at a nearby restaurant. He is there with you and does not leave, using the ordinary like breaking bread with someone at a diner, to remind you of His presence. While not every town has a restaurant like Rudy's or a place which is a touchstone for family and friends, life-giving relationships occur where two or more are gathered together. The best thing about diners is that one just needs to show up. Like the table offerings of faith in Christ, the menu will be familiar at times and will change on occasion. But the opportunity to satisfy your hunger and thirst will always be offered.

The most interesting things about seasons in our life: with each predictable moment, even more unpredictable moments occurred. My scripted and planned days turned into anything but, and as a result, offered me more gifts and lessons than I could have imagined. Memories packed away were dusted off to show with new brilliance how God was working then and is working even now. To move a fruit from farm to table first requires death of a seed, then growth in the field, then sharing of the resulting fruit onto the table of life.

Healing and restoration come when my perceived reality meets sacred revelation that God speaks in scripture and in silent moments. Moments of dying to self are shared alongside growth in the field of my life. This book becomes the offering on the banquet table for others to taste and see that the Lord is good. Like the truth that I was never meant to do life alone...and the truth that I am not alone, not by a long shot.

If you have ever known grief, ever carried hidden heartbreak, or know someone on that journey right now, this book is for you. Whether pastor, parent, or caregiver, we are all children on a journey that is not always easy. My journey of faith and hope is still a work in progress every day. Be assured – no matter how you feel - that you are not alone, that you are loved, and God is for you more than you can imagine.

A WORD IN SEASON II

When he was at the table with them, he took bread, gave thanks, broke it, and began to give it to them.

Then their eyes were opened and they recognized him,

and he disappeared from their sight. They asked each other, "Were not our hearts burning within us while he talked with us on the road and opened the Scriptures to us?"

They got up and returned at once to Jerusalem, there they found the Eleven and those with them, assembled and saying, "It is true! The Lord has risen and has appeared to Simon." Then the two told what had happened on the way, and how Jesus was recognized by them when he broke the bread.

Luke 24:30-25

Then their eyes were opened and they recognized Jesus.

Both the seen and unseen come together. Hearts burned as the Word of God pulled them closer to the reality of the divine in their midst. A new reality, a new day, a new hope is born. A hope so good that sharing the good news was a must.

My eyes were opened at the table. At the table of fellowship. At the table of connection. At the table of grace.

When I recognized the Holy across the table, breaking the bread that gives life and offering that sustenance… that is where the broken pieces of what was come together, creating something new. For when brokenness, loss, and grief are put into divine hands in trust and faith, a beautiful witness emerges.

With open eyes, I recognize the last breath on this earth does not have the final say.

With an open mind, I recognize that the past has to give way for the new that is emerging.

With an open heart, I recognize I must accept myself as a beloved child of God, and love others for they are my holy siblings.

With an open spirit, I recognize this is a broken world that needs the hope that Jesus was and is and is to come – for that is the gospel truth.

Walking on the road of life means, at times, the solid paving stones will feel like shifting sand beneath our feet. Through the landscape of ups and downs, travelers carry heartbreak and hurt of lost hope like the two on the Road to Emmaus.

And as Jesus joins us, we are oblivious to his presence. Even in the presence of the Scriptures that affirm the persistence and presence borne out of the Love of an Almighty God, we can keep looking down. Losing sight of the way. Losing sight of the truth. Losing sight of the light of hope. Hope is offered to all travelers, including you and me.

Still, the journey is necessary. Healing happens in community. Healing happens in the ordinary. Healing happens in divine interruptions. Healing happens at the table, where we break bread with another.

This book tells of one such season of broken hopes and hearts in need of hope, a hope found in the most unexpected of places. Your season may look different than mine, yet still leaving you with the words "I had hoped…" Hope is waiting for you to join the journey toward healing through love and grace. All you need to do is find an open table and begin your journey.

Take heart, weary one. You are not alone.

Resources

Study Guide Outline

(for small groups, retreats, or spiritual formation courses)

Breakfast at Rudy's – Group Study Guide (6–10 Weeks)

Overview: This guide is designed for small groups, pastoral cohorts, grief support circles, or spiritual formation classes. Each session includes key themes, reflection questions, and spiritual practices.

Session 1: Breaking Bread with Grief

Read: *Landscape of Loss* + Letter 1
Themes: Naming loss, holding sacred space, showing up at the table
Questions:

- What has been your "Rudy's"—a place where grief or grace found you?

- How have you felt "in pieces" this season?

Practice: Write your brief letter to God naming a loss.

Session 2: Appetite and Absence

Read: *Loss of Appetite*
Themes: Grieving a loved one, reclaiming joy in small ways
Questions:

- What do you hunger for in this season—emotionally, spiritually?

- What memory of someone you've lost brings both tears and laughter?

Practice: Share a story of a lost loved one and the food or space that reminds you of them.

Session 3: Burnout, Ministry, and Being Enough

Read: *On My Own*
Themes: Loneliness in leadership, spiritual weariness, caregiving
Questions:

- What has your experience of ministry or caregiving cost you emotionally?

- Where have you felt God most silent—and most near?

Practice: Sabbath practice: take 1 hour this week to rest with no productivity goals.

Session 4: Fast Faith vs. Slow Presence

Read: *Fast Food Faith*
Themes: Rushing spirituality, transactional prayer, sacred presence
Questions:

- Where in your life have you substituted speed for presence?

- What's your rhythm of "sitting with God"?

Practice: 10-minute silent prayer or Lectio Divina on Psalm 34:8.

Session 5: Remembering, Reclaiming

Read: *Blue Ribbon Special*
Themes: Legacy, storytelling, healing over time
Questions:

- What stories from your family shape your view of God?

- Who do you need to "see again" with fresh eyes?

Practice: Write a 1-paragraph spiritual memory about someone who shaped your faith.

Session 6: From Loss to Life Again

Read: *Landscape of Hope* + your favorite blessing
Themes: Hope after heartbreak, divine interruptions, daily resurrection
Questions:

- How has your idea of "healing" changed over time?

- What blessing from the book would you give someone you love today?

Practice: Write a blessing for someone going through grief or burnout.

Closing Thought: God speaks in the stillness, and His love is woven into every memory, every silence, and every sunrise. May this time be an opportunity for healing, rekindling, and drawing close to the heart of God. Amen.

Letter Writing Guide to God

1. Pray and talk to God before you write and listen to the Spirit leading you.

Likewise, the Spirit also helps in our weaknesses. For we do not know what we should pray for as we ought, but the Spirit Himself makes intercession for us with groanings which cannot be uttered. Romans 8:26

2. Begin where you are

In the beginning was the Word, and the Word was with God, and the Word was God. He was with God in the beginning. Through him all things were made; without him nothing was made that has been made." John 1:1-3

    a. If you are thankful to God but don't understand, then state that.
    b. If you are upset with God, be honest.
    c. If you cannot even express gratitude, be thankful for the opportunity to put your thoughts in writing
    d. Always take as much time listening as you do for venting, sharing, or talking.

3. Share where you have been and where you are

Hear my cry, O God; attend unto my prayer. From the end of the earth will I cry unto thee, when my heart is overwhelmed: lead me to the rock that is higher than I. Psalm 61:1-2

a. Share your circumstances, past and present. Think of it as the all-knowing God is asking you to share your days and experiences; not because He doesn't know them, but because those who love you want to hear from you.
b. Share concerns about the future, if applicable
c. Include what you are feeling, hearing, seeing, or the lack thereof
d. Always take time to listen for God's response, in Scripture or Spirit.

4. **Ask forgiveness for any mistakes or misunderstandings on your part**

The sacrifice you desire is a broken spirit. You will not reject a broken and repentant heart, O God. Ps. 51:17

For you, O Lord, are good and forgiving, abounding in steadfast love to all who call upon you. Ps. 86:5

If others are involved in your brokenness, offer mercy the best you can, followed by grace and forgiveness:

a. Forgiveness does not condone an action but gives it to God and lets go of perceived control of the situation
b. Take time to receive forgiveness for yourself, and to let go of any unforgiveness you are holding onto.

5. **Seek God's guidance and strength in doing the hard work ahead, such as forgiveness, confession, repentance, humility, etc.**

    For in my inner being I delight in God's law; but I see another law at work in me, waging war against the law of my mind and making me a prisoner of the law of sin at work within me. What a wretched man I am! Who will rescue me from this body that is subject to death? Thanks be to God, who delivers me through Jesus Christ our Lord! Romans 7:22-25a.
    Be willing to write what you hear the Spirit saying in scripture, song, or some other manner.

6. Find peace in God's Word for you

I have told you these things so that in me you may have peace. In this world, you will have trouble. But take heart! I have overcome the world. John 16:33

Be still and know God is present with you in mind, body, and spirit. Share in your journal or with someone else, what does God say, and how does that make you feel at peace instead of pieces?

7. Thank God for listening and for who God is

"I love the Lord, for he heard my voice; he heard my cry for mercy. Because he turned his ear to me, I will call on him as long as I live." Ps. 116:1-2

Give thanks with rejoicing. Joy is found in the morning as a new day starts, and this letter is meant to be an inauguration of a new day for you.

8. **Commit to staying in two-way communication (sharing and listening) in the days ahead.**

For it is [not your strength, but it is] God who is effectively at work in you, both to will and to work [that is, strengthening, energizing, and creating in you the longing and the ability to fulfill your purpose] for His good pleasure. Phil. 2:13 AMP

Schedule time with God every day when you can be quiet. And when interruptions come, welcome the interruption as a moment to lean in and listen more closely to what the Holy Spirit is saying/doing in that moment. Even if interruptions are uncomfortable, painful, or unwanted, keep showing up and putting your trust in God.

9. **Sign off with love for the God who loves you**

I love you, Lord, my strength. The Lord is my rock, my fortress, and my deliverer; my God is my rock, in whom I take refuge, my shield and horn of my salvation, my stronghold. Psalm 19:1-2

Do not worry about grammar or syntax. Just write from your heart.

*After writing the letter, you can read it out loud as a prayer to God, or you may share it with others. You may choose to keep the letter in a sacred space (Bible, journal, etc.) or you may choose to burn it as an offering.

Making Prayer Personal

When feeling overwhelmed, prayer can recenter the focus from problems one has to the promises God gives. God is always faithful, even when we are not.

One way to begin a prayer is to pray scripture. The psalms are a good place to start because they contain the deep emotions we feel in life. Another option is verses such as the beloved verse of John 3:16-17.

For God so loved the world that he gave his one and only Son, that whoever believes in him shall not perish but have eternal life. For God did not send his Son into the world to condemn the world, but to save the world through him. John 3:16-17

In this case, inserting your name where "the world" is shown is a way to start praying scripture from a personal perspective. * *Note: A little revision on sentence structure helps make this pattern of prayer flow while maintaining the purpose and intent of the passage.*

> For God so loved __**ME**__ that he gave his one and only Son, that (whosoever, like) ___**ME**___ believes in him shall not perish but have eternal life. For God did not send his Son to condemn ___**ME**___, but to save ___**ME**___ through him. John 3:16-17

Another way to commit or reaffirm your commitment to Christ is through the Sinner's Prayer. If you have never received Christ as your Lord and Savior, or if you have wandered away from Him, you will never know true peace and love without giving your life to Christ. Even if you are a disciple of Jesus, daily recommitting to him through this prayer is a way to remind yourself of where you stand and who Jesus is to you. While many variations of this prayer are offered, the prayer includes 4 main parts as one talks to God: I admit, I believe, I commit and fill me with your Spirit. An example is listed below:

Jesus, I admit I am a sinner, and I can't save myself.

I believe that you love me. I believe you died for me, forgive my sins.

I believe that you rose again to offer me new life now and for all eternity.

I commit my life to you and invite you into my heart.

To be my Lord and lead me every day. To be my savior from sin.

Fill me with your Holy Spirit and help me live a life pleasing to you.

In Jesus' name, I pray. Amen.

Bottom line: No matter where you have been or where you are now in your life. No matter how low or lost or hurting you may feel. Honest and fervent prayer is the key to living life in abundance. Jesus reminds us there will be trials and troubles while we live on this earth, but they don't have the final say – Jesus does for He has overcome the world. Relationship with God begins with prayer and is strengthened by prayer. Prayer – both speaking and listening - is essential to be a believer in Jesus Christ and follow in the life He offers.

Gratitudes

First and foremost, I thank Jesus Christ, my Lord and Savior, who by His Spirit is always shaping and molding me into more for His glory. My life has been full of good choices and bad choices, opportunities seized and opportunities missed. Still, God has never given up on me and steadily goes before me, behind me, and by my side every step of the journey until I see Him face to face in glory. And in His love, He has put wonderful people in my life to also join me on the journey.

To my beloved family, you are such a blessing to me. I am thankful for my parents, Joseph and Bettye Prettyman, who gave me their legacy of love and learning that extended through their lives on this earth and beyond. Both have now died, but live in my heart always. I am thankful for my husband, Glenn, and my sons – Randy, Robby, Patrick, and Joe – who encouraged me, especially in my difficult seasons. Patrick, you were always willing to read my first, second, and twentieth drafts, and for that I am eternally grateful. And with this book, Randy – Robby – Joe joined the "how does this sound?" team.

To my faithful friends who have read pages, shared wisdom, and offered brotherly and sisterly love on the journey that became this book, I thank you. A special thank you to the spiritual director who saw the flowing river beyond my walls.

To my fellow servants in ministry, your patience and partnership with me over the years have helped keep me on the path God has called me to. To my dear siblings from different denominations and faith expressions, you have made more of a difference in my personal and pastoral life than you will ever know.

Finally, to my colleagues in the publishing world who know so much more about the physical making of a book than me, thank you for your patience and perseverance. Thank you to Red M., whose persistence reminded me not to quit, Kim B., whose vision was bigger than mine, and to the team that helped make the publication of this final manuscript come together.

And God, you always get my first, my last, and my deepest gratitude for meeting me wherever I find myself and never leaving me there unchanged.

About the Author

Rebecca Collison is a teacher-turned-pastor who has served many small and medium-sized churches located on the East Coast of the United States. While a parent and asn advocate for people who have been labeled as having disabilities, she has always had a heart for all of God's children in search of healing and wholeness. This book was not an academic endeavor, but a personal and heartfelt sharing that emerged during a three-month spiritual renewal retreat, a retreat to reset her focus after losses and challenges physically, emotionally, and spiritually wore her down over 20 years of ministry, specifically losses in the previous four years.

This book was written to reach out to others, pastors included, who feel overwhelmed and broken by the times and trials of the world. Grief, burnout, and brokenness give way to hope, mercy, and grace in the healing process. This personal transformation is offered in her honest sharing of her journey, punctuated by reflections, Scripture references, and blessings along the way.

www.ingramcontent.com/pod-product-compliance
Lightning Source LLC
Chambersburg PA
CBHW021653120626
46545CB00002B/846